Contents

Preface

About the Author

Chapter 1: The Living World

Biodiversity

Types of Biodiversity

Three Domains of Life

Taxonomy

Systematics

Concept of Species

Taxonomical Hierarchy

Binomial Nomenclature

Museums

Zoological Parks

Herbaria

Botanical Garden

Chapter 2: Biodiversity and It's Conservation

Biodiversity Patterns

Importance

Loss of Biodiversity

Biodiversity Conservation

Hotspots

Endangered Species

Red Data Book

Biosphere Reserves

National Parks

Sanctuaries

Ramsar Sites

Preface

Welcome to biodiversity. Biodiversity is the variety and variability of life on Earth. Without biodiversity, we do not live. Human population is increasing day by day and destroy the biodiversity for various needed. Biodiversity refers to varieties of living things i.e plants, animals, bacteria, and fungi. Biodiversity provides food, water, timber, fiber, and genetic resources.

The book comprises two chapters. The first chapter gives definition of living, biodiversity, taxonomy, systematics, and species. The first chapter also gives types of biodiversity, three domains of life, taxonomical hierarchy, binomial nomenclature, role of botanical garden, importance of herbaria. The second chapter deals with the biodiversity patterns, importance of biodiversity, and loss of biodiversity. The second chapter also discusses biodiversity conservation, hotspots, endangered species, red data book, biosphere reserves, national park, sanctuaries, and ramsar sites.

Thanks to my family. Finally, thanks to all the friends for their continue support and encouragement.

Chapter 1

The Living World

What is living?

Living means alive. Living things are made up of cells, that requires energy. Living things can grow, move, reproduce, and respond to their environment.

Biodiversity:

Biodiversity is the variety and variability of life on Earth. Without biodiversity, we do not live. Human population is increasing day by day and destroy the biodiversity for various needed. Biodiversity refers to varieties of living things i.e plants, animals, bacteria, and fungi. Biodiversity provides food, water, timber, fiber, and genetic resources.

According to U.S. Office of Technology Assessment (1987), biological diversity is "the variety and variability among living organisms and the ecological complexes in which they occur."

Types of Biodiversity:

There are three types of biodiversity-

i. Genetic Diversity: Genetic diversity is refers to the diversity in the genetic makeup within the species.

ii. Species Diversity: Species diversity is the number of species live in a particular location. Species diversity refers to the number of species found in ecological community.

iii. Ecosystem Diversity: Ecosystem diversity is the variety of ecosystem within an area. Ecosystem diversity includes both biotic and abiotic component.

Three Domains of Life:

The evolutionary relationships between organisms are subject of phylogeny. The phylogenetic tree showing the evolutionary relationships among various biological species or other entities. All life on earth is part of a single phylogenetic tree, indicating common ancestry. The microbial world has three main cell lineages which are thought to have evolved from a single progenitor. The lineages are formally knowm as Domain. The three domain system, proposed by woese and others, is an evolutionary model of phylogeny based on differences in the sequences of nucleotides in the cell's rRNA, as well as cell's membrane lipid structure and its sensitivity to antibiotics. The phylogenetic tree consist of three domains of organisms the Bacteria and the Archae, cells of which are prokaryotic, and the Eukarya (eukaryotes). Eukaryota and Archaea are more closely related to each other than Bacteria (based on Caliver-Smith's theory of bacterial evolution). Each of these three domains contains unique rRNA. rRNA is the RNA component of the ribosome, which are essential for protein synthesis in all living organisms.

Property	Archaea	Bacteria	Eukarya
Cell membrane	Glycerol diether linked lipids.	Ester-linked phospholipids and hopanoids.	Ester-linked phospholipids and sterols.
Cell wall	Glycoproteins	Peptidoglycan	Various structure
Endoplasmic reticulum	Absent	Absent	Present
Golgi apparatus	Absent	Absent	Present
Lysosome	Absent	Absent	Present
Mitochondria	Absent	Absent	Present
Chloroplast	Absent	Absent	Present
Nucleolus	Absent	Absent	Present
RNA polymerase	Many	One	Many
Toxin	Sensitive to diphtheria toxin	Resistant to diphtheria toxin	Sensitive to diphtheria toxin
Histone	Absent	Present	Present

Taxonomy:

Taxonomy is come from two greek words taxis meaning arrangement and nomia meaning method. Taxonomy encompasses the description, identification, nomenclature, and classification. Taxonomy includes all plants, animals, and microorganisms.

The term taxonomy was first coined by A.P de Candolle in 1813.

Systematics:

The term taxonomy and systematics have been interchangeably used in the past. According to Radford in 1986, systematics is the study of phenotypic, genetic, and phylogenetic relationship among taxa.

Objectives of Systematics:

i. Understading of the evolutionary process.

ii. To establish a suitable method for identification, nomenclature, and description of plant taxa.

Concept of Species:

Species: Species is refers to the group of living organisms that can be produce fertile offsprings.

Taxonomical Hierarchy:

Taxonomical Hierarchy is the sequence of arranging various organisms into the biological classifications.

Here, Kingdom is the highest rank and species is the lowest rank in this hierarchy.

Binomial Nomenclature:

Binomial nomenclature is the system of naming species or organisms of living things.

All living organisms like plants, animals, birds, and some microorganisms have their own scientific names. For e.g.

- The Scientific name of the Mango is represented as Mangiferaindica. Here, Mangifera represents the genus and indica represents the particular species.
- The Scientific name of the Leopard is presented as Pantherapardus . Here Panthera represents the genus and pardus represents a species.

Codes:

ICZN- International Code of Zoological Nomenclature.

ICNafp- International Code of Nomenclature for algae, fungi, and plants.

ICNB- International Code of Nomenclature of Bacteria.

ICTV- International Committe on Taxonomy of viruses.

Botanical Gardens:

Botanical Gardens are the institutions. Botanical Gardens maintain the collection, preservation, and cultivation of different varieties of plants.

Role of Botanical Garden:

i. Botanical Garden provide information of local flora.

ii. Botanical Garden supply seeds, flowers, fruits, and materials.

iii. Many peoples are visit botanical garden. Botanical gardens attract many peoples.

iv. Botanical Garden conserve rare species.

v. Botanical Garden provide education to teacher, students, and scientists.

List of botanical gardens in India:

i. Indian Botanic Garden, Kolkata.

ii. Botanical Garden, Sharangpur.

iii. Lalbagh Botanical Garden, Bangalore.

iv. Government Botanical Gardens, Ooty.

v. Tropical Botanic Garden, Trivandrum.

vi. Lloyd Botanic Garden, Darjeeling.

vii. Jawaharlal Nehru Botanical Garden, Gangtok.

viii. National Botanic Garden, Lucknow.

ix. Assam State Botanical Garden, Guwahati.

x. SemmozhiPoonga Botanical Garden, Chennai.

xi. Pilikula Botanical Garden, Mangalore.

xii. Reddy Botanical Garden, Hyderabad.

xiii. Botanical Garden, Chandigarh.

xiv. Waghai Botanical Garden, Saputara.

xv. Nehru Memorial Botanical Garden, Srinagar.

Table 1. List of Botanical Gardens in various country:

Country	Botanical Gardens	Formed/Founded
Abkhazia	Sukhumi Botanical Garden	1840
Argentina	Administraction de parquesNacionales	September 30, 1934
Argentina	Buenos Aires Botanical Garden	September 7, 1898
Armenia	Yerevan Botanical Garden	1935
Armenia	IjevanDendropark	1962
Armenia	Sevan Botanical Garden	1944
Armenia	StepanavanDendropark	1931
Austrilia	Austrilian National Botanic Garden	1949
Austrilia	Lindsay Pryor National Arboretum	
Austrilia	Westbourne Woods	
Austrilia	Albury Botanic Gardens	1877
Austrilia	Auburn Botanical Garden	1977
Austrilia	Austrilian Inland Botanical Garden	1989
Austria	Innsbruck University Botanic Garden	1911
Austria	Botanical Garden of the University of Vienna	
Bangladesh	National Botanical Garden of Bangladesh	1961
Bangladesh	Balda Garden	1909
Barbados	Andromeda Botanical Gardens	
Barbados	Hunte's Gardens	
Belgium	Antwerp Botanic Garden	1825
Belgium	Botanical Garden of Brussels	1826
Belgium	Arboretum Kalmthout	1856
Belgium	HortusBotanicusLovaniensis	1837
Belize	Belize Botanic Gardens	
Bermuda	Bermuda Botanical GardenS	
Botswana	National Botanical Garden	November 2, 2007
Singapore	Singapore Botanic Gardens	1859
Singapore	Gardens by the Bay	29 June 2012
Vietnam	Saigon zoo and Botanical Garden	1865
United States	Alaska Botanical Garden	1993
United States	Georgeson Botanical Garden	

Uganda	National Botanical Gardens	1898
Zimbabwe	National Botanical Garden of Zimbabwe	1902
India	Sanjay Gandhi JaivikUdyan, Patna	1973
India	National Cactus and Succulent Botanical Garden and Research Centre, Haryana	1987

Herbarium:

Herbarium is the collection of plant materials in any place. After collection of plants, plants should be dried, pressed and arranged according to the classifications system. Herbarium may be mounted on a sheet of paper.

Luca Ghini is the creator of the art of herbarium.

Figure 4. Herbarium specimens of Crotalaria juncea (Photo Credit: AnupamRajak)

Importance of Herbaria:

i. Herbaria are used in taxonomic research.

ii. Provide data for floristic studies.

iii. Herbarium provide knowledge about the flora.

Table 1. List of Herbaria:

Name	No. Specimens
Museum of Natural History, Paris	6.5 million
Komarov Botanical Institute, Leningrad	Over 5 million
Conservatory and Botanical Garden, Geneva	5 million
Combined Herbaria, Harvard University, Cambridge	4.5 million
New York Botanical Garden, Bronx	4.3 million
U.S National Herbarium	4.1 million
British Museum of Natural History, London	4 million

Museum:

Museums are the institution where educational materials are showing to the public. Museums preserve selected objected. Museums give education to the public.

Figure 5. Museums (Photo Credit: Pixabay)

Table 1. List of Museums

Country	Name of the Museum	Established
India	BhagwanMahabir Government Museum	1982
India	Victoria Jubilee Museum	1887
India	Assam State Museum	1940
India	Goa Chitra Museum	2010
India	Swaminarayan Museum	2011
United States	Alabama Administrative office of Courts Museum Area	1994

Zoological Park:

Zoological parks are the institution where living organisms are kept and exhibited to the public.

Figure 6. Nandankanan Zoological Park (Photo Credit: Flickr)

List of Zoological parks-

i. National Zoological Parks, New Delhi.

ii. Rajiv Gandhi Zoological Park, Maharashtra.

iii. Nandankanan Zoological Park, Odisha.

iv. Indira Gandhi Zoological Park, Andhra Pradesh.

v. Padmaja Naidu Himalayan Zoological Park, Darjeeling.

Chapter 2

Biodiversity and It's Conservation

Biodiversity Patterns:

Darwin noticed three patterns of biological diversity-

i. Species vary globally.

ii. Species vary locally.

and iii. Species vary over time.

Ecologists have studied various patterns of species biodiversity-

i. Latitudinal Gradients- When, we move from equators, towards the poles, the diversity of species is decreases.

ii. Species-Area Realationships: Species-Area Relationships are the number of species found on the Earth.

Log scale represents the following equations-

$logS = logC + zlogA$

Here, S= species richness.

A= Area

Z= Slope of the line.

C= Y- intercept

Importance of Biodiversity:

Biodiversity provides us to oxygen, air, water, timber, and many other thing. Biodiversity provides large number of plant species. Biodiversity provides food, medicine, and drugs.

Loss of Biodiversity:

Biodiversity loss is the extinction of loss of species in a certain habitat.

i. Deforestation: We cut down the trees for various purpouses. We destroy the ecosystems.

i. Overexploitation: Hunting and poaching of the species is the major causes for the loss of biodiversity.

iii. Pollution: Pollution is the major causes for the loss of biodiversity. We know that, plastics are dumped into the ocean surface. Plastics are polluted our environment and earth ecosystems.

Burnning of fossils fuels are polluted into the atmosphere.

iv. Lastly say, climate change and global warming is the major causes for the loss of biodiversity.

Biodiversity Conservation:

Biodiversity conservation is the protection, conservation of biodiversity for sustainable development.

Types of Conservation:

In Situ Conservation:

In situ conservation is the conservation of plant or animal species. In situ conservation includes National Park and sanctuaries, biosphere reserves, nature reserves, reserve and protected forests, preservation plots, reserved forests etc.

Ex Situ Conservation:

Ex situ conservation is the conservation of living organisms. Ex situ conservation includes zoological and botanical parks.

Figure 4. Ex situ conservation (Photo Credit: Wikimedia Commons/EanPaerKarthik / CC BY-SA (https://creativecommons.org/licenses/by-sa/4.0))

Hotspots:

Biodiversity hotspots are the contain at least 1500 species of vascular plants nowhere else on Earth. Biodiversity includes both flora and fauna. Biodiversity hotspots regions are particularly rich in endemic, rare, and threatened species.

Biodiversity hotspots concept was first introduced by Norman Myers in 1988.

There are at least 34 biodiversity hotspots regions around the World-

Africa

1. Eastern Afro-Montane

2. The Guinean forests of Western Africa

3. Horn of Africa

4. Madagascar and the Indian Ocean Islands

5. Maputoland, Podoland, Albany hotspot

6. Succulent Karou

7. East Malanesian islands

8. South Africa's Cape floristic hotspot

9. Coastal forests of Eastern Africa

Terrestrial Biomes of the World

Asia and Australia

1. Himalayan hotspot

2. The Eastern Himalayas

3. Japan biodiversity hotspot

4. Mountains of South-West China

5. New Caledonia

6. New Zealand biodiversity hotspot

7. Philippine biodiversity hotspot

8. Western Sunda (Indonesia, Malas and Brunei)

9. Wallace (Eastern Indonesia)

10. The Western Ghats of India and Islands of Sri Lanka

11. Polynesia and Micronesian Islands Complex including Hawaii

12. South-Western Australia

North and Central America

1. California Floristic Province

2. Caribbean islands hotspot

3. Modrean pine-oak wood lands of the USA and Mexico border

4. The Mesoamerican forests

Aquatic Biomes of the World

South America

1. Brazil's Cerrado

2. Chilean winter rainfall (Valdivian) Forests

3. Tumbes-Choco-Magdalena

4. Tropical Andes

5. Atlantic forest

Europe and Central Asia

1. Caucasus region

2. Iran-Anatolia region

3. The Mediterranean basin and its Eastern Coastal region

4. Mountains of Central Asia

Endangered Species:

Endangered species is a species i.e very extinct in near future.

Some endangered species are listed below-

i. Orangutan (Pongopygmaeus).

ii. Tasmanian devil (Sarcophilusharrisii).

iii. Gorilla (Gorilla beringei).

iv. Snow leopard (Panthera uncial).

v. Sea otter (Enhydralutris).

vi. Asian elephant (Elephasmaximus).

vii. Blue whale (Balaenopteramusculus).

viii. Whooping Crane (Grusamericana).

ix. Tiger (Panthera Tigris).

x. Giant Panda (Ailuropodamelanoleuca).

Red Data Book:

The Red data book is a public document which is originate from Russia. The Red data book is recording endangered species of plants, animals, and fungi.

The Red data book contains various colour represents coded information sheets i.e listed below-

Red- endangered.

Black- Species, which are confirmed to be extinct.

Amber- Vulnerable.

White- Rare.

Green- endangered but their numbers have started to recover.

Grey- Vulnerable, endangered, or rare.

Biosphere Reserves:

Biosphere reserves are nominated by national government. Biosphere reserves consists of three zones i.e core area, buffer zone, and transition area.

According to UNESCO, Biosphere reserves are areas comprising terrestrial, marine, and coastal ecosystems that are recognised by it's the Man and the Biosphere Programme (MAB, 1971).

List of Biosphere Reserves in India-

i. Nilgiri Biosphere Reserves, Tamil Nadu, Kerala, and Karnataka.

ii. Gulf of Mannar, Tamil Nadu.

iii. Sundarbans Biosphere Reserves, West Bengal.

iv. Nanda Devi Biosphere Reserves, Uttarakhand.

v. Nokrek Biosphere Reserves, Meghalaya.

vi. Pachmarhi Biosphere Reserves, Madhya Pradesh.

vii. Simlipal Biosphere Reserves, Odisha.

viii. Great Nicobar Biosphere Reserves, Andaman and Nicobar Islands.

ix. Agasthyamalai Biosphere Reserves, Kerala and Tamil Nadu.

National Park:

National parks are recognized by national government. National parks are protect the flora and fauna. National parks is an area where reserved of the wildlife and biodiversity.

The first national park is Yellowstone National Park in 1872. The largest national park is Hemis National Park, Jammu and Kashmir.

List of National park in India:

i. Campbell Bay National Park, Andaman and Nicobar Islands.

ii. Rajiv Gandhi National Park, Andhra Pradesh.

iii. NamdaphaNational Park, Arunachal Pradesh.

Iv, Kaziranga National Park, Assam.

v. Valmiki National Park, Bihar.

vi. Guru Ghasidas National Park, Chattisgarh.

vii. Mollem National Park, Goa.

viii. Vansda National Park, Gujarat.

ix. Kalesar National Park, Haryana.

x. Pin Valley National Park, Himachal Pradesh.

xi. Dachigam National Park, Jammu and Kashmir.

xii. Betla National Park, Jharkhand.

xiii. Bandipur National Park, Karnataka.

xiv. Silent Valley National Park, Kerala.

xv. Kanha National Park, Madhya Pradesh.

xvi. Pench National Park, Maharashtra.

xvii. Keibul-Lamjao National Park, Manipur.

xviii. Nokrek Ridge National Park, Meghalaya.

xx. Buxa National Park, West Bengal.

Sanctuaries:

Sanctuaries are an area where animals are protected from hunting.

Figure 9.Bharatpur Bird Sanctuary (Photo Credit: Pixabay)

List of Wildlife Sanctuaries in India:

i. AriakIsland, Andaman and Nicobar Island.

ii. Barren Island, Andaman and Nicobar Island.

iii. Kolleru Bird Sanctuary, Andhra Pradesh.

iv. Pulicat Lake Bird Sanctuary, Andhra Pradesh.

v. Pakke Tiger Reserve, Arunachal Pradesh.

vi. Bornadi Wildlife Sanctuary, Assam.

vii. Bhimbandh Wildlife Sanctuary, Bihar.

Viii. Udaypur Wildlife Sanctuary, Bihar.

ix. Achanakmar Wildlife Sanctuary, Chattisgarh.

x. Dadra and Nagar Haveli Wildlife Sanctuary, Dadra Nagar Haveli.

xi. Salim Ali Bird Sanctuary, Goa.

xii. Bhindawas Wildlife Sanctuary, Haryana.

xiii. Gir Wildlife Sanctuary, Gujarat.

xiv. Dalma Wildlife Sanctuary, Jharkhand.

xv. Neyyar Wildlife Sanctuary, Kerala.

xvi. Great Indian Bustard Sanctuary, Maharastra.

xvii. Kotgarh Wildlife Sanctuary, Odisha.

xviii. Maenam Wildlife Sanctuary, Sikkim.

xx. kumbhalgarh Wildlife Sanctuary, Rajasthan.

xxi. Vedanthangal Bird Sanctuary, Tamil Nadu.

xxii. Pocharam Wildlife Sanctuary, Telangana.

xxiii. Gumti Wildlife Sanctuary, Tripura.

xxiv. Hastinapur Wildlife Sanctuary, Uttar Pradesh.

xxv. Binsar Wildlife Sanctuary, Uttarakhand.

xxvi. Chapramari Wildlife Sanctuary, West Bengal.

Ramsar Sites:

The Convention on wetlands are also known as Ramsar Convention. Ramsar Convention was established in 1971. Ramsar is a city of Iran.

Figure 9. Animal Lake Japan (Photo Credit: Pixabay)

International Organizations Partners:

There are six organizations are listed below-

- Birdlife International.
- International Union for Conservation of Nature (IUCN).
- International Water Management Institute (IWMI).
- Wetlands International.
- WWF International.
- Wildfowl and wetlands trust (WWT).

List of Ramsar Sites-

i. Ashtamudi wetland, Kerala.

ii. Bhitarkanika Mangroves, Odisha.

iii. Bhoj wetland, Madhya Pradesh.

iv. Chandra Taal, Himachal Pradesh.

v. Chilka Lake, Odisha.

vi. DeeporBeel, Assam.

vii. East Kolkata Wetlands, West Bengal.

viii. Harike Wetlands, Punjab.

ix. Hokera Wetland, Jammu and Kashmir.

x. Kanjli Wetland, Punjab.

xi. Kolleru Lake, Andhra Pradesh.

xii. Loktak Lake, Manipur.

xiii. Pong Dam Lake, Himachal Pradesh.

References:

1. htttps://byjus.com/biology/what-is-living/
2. https://www.theguardian.com/news/2018/mar/12/what-is-biodiversity-and-why-does-it-matter-to-us
3. https://www.nationalgeographic.org/encyclopedia/biodiversity/
4. https://www.vedantu.com/biology/biodiversity
5. https://en.m.wikipedia.org/wiki/Biodiversity
6. https://www.safeopedia.com/definition/2992/species-diversity
7. https://socratic.org/questions/what-is-ecosystem-diversity
8. http://www.coastalwiki.org/wiki/Ecosystem_diversity
9. https://www.cbd.int/gti/taxonomy.shtml
10. https://byjus.com/biology/concept-of-species/
11. https://byjus.com/biology/binomial-nomenclature/
12. https://en.m.wikipedia.org/wiki/Botanical_garden
13. http://www.biologydiscussion.com/articles/importance-of-botanical-gardens/6522
14. https://www.padeepz.net/herbaria-and-uses-importance-of-herbaria/
15. https://en.m.wikipedia.org/wiki/Museum
16. https://www.takshilalearning.com/class-12-biology-patterns-of-biodiversity/
17. https://soe.environment.gov.au/theme/biodiversity/topic/2016/importance-biodiversity
18. http://ete.cet.edu/gcc/?/biodiversity_importance/
19. https://openoregon.pressbooks.pub/envirobiology/chapter/123/
20. https://mashable.com/2015/05/23/biodiversity-threats/
21. https://www.greenfacts.org/en/biodiversity/l-3/4-causes-desertification.htm
22. https://byjus.com/biology/biodiversity-conservation/
23. https://swww.cepf.net/our-work/biodiversity-hotspots/hotspots-defined
24. https://m.jagranjosh.com/general-knowledge/biodiversity-hotspots-of-the-world-1523356211-1

25. https://www.britannica.com/list/10-of-the-most-famous-endangered-species

26. https://www.unescomedcenter.org/en/biosphere-reserves

27. https://en.m.wikipedia.org/wiki/Biosphere_reserves_of_India

28. https://simple.m.wikipedia.org/wiki/National_park

29. https://www.careerpower.in/national-parks-india.html

30. https://en.m.wikipedia.org/wiki/List_of_wildlife_sanctuaries_of_India

31. https://en.m.wikipedia.org/wiki/Ramsar_Convention

32. https://en.m.wikipedia.org/wiki/List_of_Ramsar_sites_in_India

33. https://en.m.wikipedia.org/wiki/Species%E2%80%93area_relationship

Thank You